# KiCad - CvPcb Reference Manual

A catalogue record for this book is available from the Hong Kong Public Libraries.

Published in Hong Kong by Samurai Media Limited.

Email: info@samuraimedia.org

ISBN 978-988-8381-88-3

# Contents

# 1 Introduction to CvPcb

CvPcb is a tool that allows you to associate components in your schematic to component footprints used when lay out the printed circuit board. This association is added to the net list file created by the schematic capture progr Eeschema.

The net list file generated by Eeschema specifies which printed circuit board footprint is associated with each compor in the schematic only when the footprint field of the component is initialized.

This is the case when component footprints are associated during schematic capture by setting the component footprint field, or it is set in the schematic library when loading the symbol.

CvPcb provides a convenient method of associating footprints to components during schematic capture. It prov footprint list filtering, footprint viewing, and 3D component model viewing to help ensure the correct footprin associated with each component.

Components can be assigned to their corresponding footprints manually or automatically by creating equivalence : (.equ files). Equivalence files are lookup tables associating each component with it' s footprint.

This interactive approach is simpler and less error prone than directly associating the footprints in the schema editor.

CvPcb allows you to see the list of available footprints and to display them on the screen to ensure you are associal the correct footprint.

**It can be run only from Eeschema**, from the top toolbar, either when Eeschema is started from the KiCad pro manager or when Eeschema is started as a stand alone application.

Running CvPcb from Eeschema lauched from the KiCad Manager is generally better because:

- Cvpcb needs the project config file to know the footprint libraries to load.

- Cvpcb initializes the components footprint fields of the current schematic project. This is possible only if project file is in the same path as the open schematic.

Lauching CvPcb from an Eeschema launched from the KiCad manager assures automatically all this.

---

**Warning**

You actually **can** launch CvPcb from a stand alone Eeschema session though, but please note that any schem opened that does not have a project file in the same path may be missing components due to missing libr which will not show up in CvPcb. If there is no fp-lib-table file in the same path as the open schematic, no pro specific footprint libraries will be available either.

---

# 2 CvPcb Features

## 2.1 Manual or Automatic Association

CvPcb allows for interactive assignment (manual) as well as automatic assignment via equivalence files.

---

# 3  Invoking CvPcb

**CvPcb is only invoked from the schematic capture program Eeschema**, by the tool:

Eeschema automatically passes the correct data (component list and footprints) to CvPcb. There is no update to (unless some new components are not yet annotated), just run Cvpcb.

# 4  CvPcb Commands

## 4.1  Main Screen

The image below shows the main window of CvPcb.

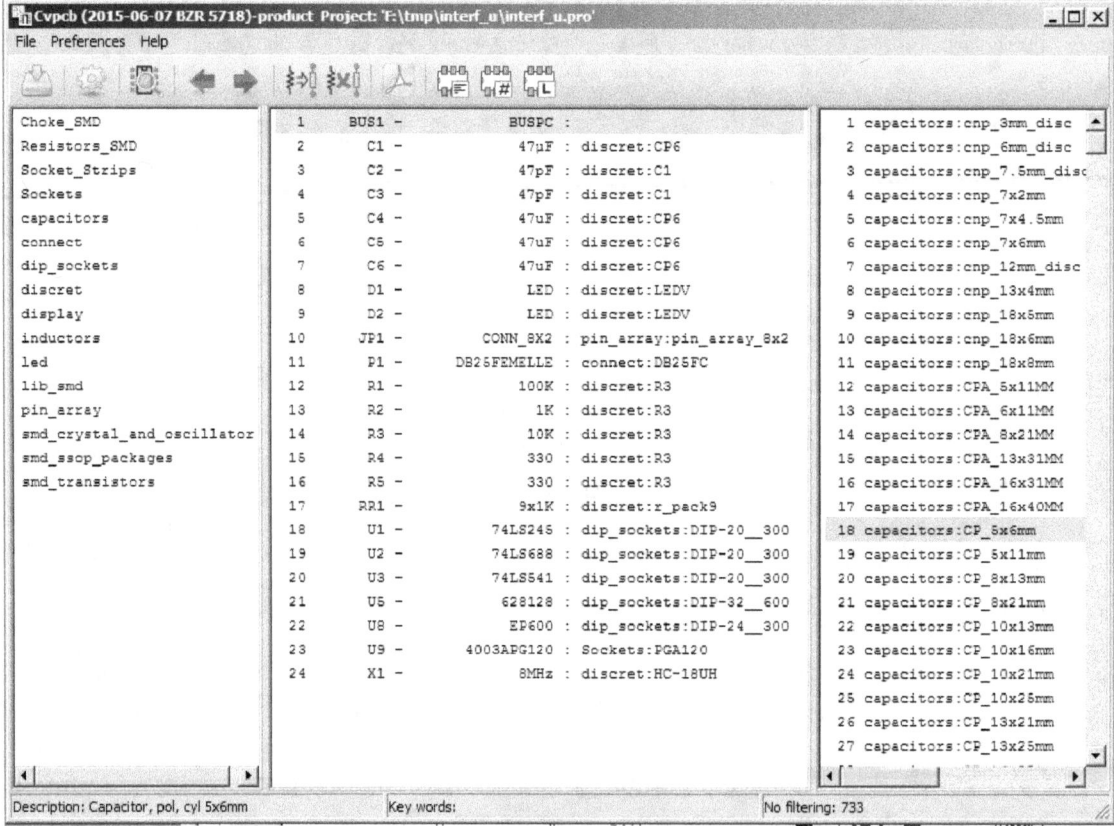

The left pane contains the list of available footprint library file names associated with the project. The center p contains the list of components loaded from the net list file. The right pane contains the list of available footpr loaded from the project footprint libraries. The component pane will be empty if no netlist file has been loaded the footprint pane can be also empty if no footprint libraries are found.

## 4.2  Main Window Toolbar

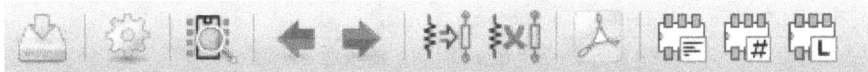

The top toolbar allows for easy access to the following commands:

| | |
|---|---|
| | Transfer the current footprint association to Eeschema (this is the content of footprint fields). |
| | Invoke the CvPcb configuration menu. |
| | Display the footprint of the component selected in the footprint window. |
| | Automatically select the previous component in the list without a footprint association. |
| | Automatically select the next component in the list without a footprint association. |
| | Automatically associate footprints with components starting using an equivalence file. |
| | Delete all footprint assignments. |
| | Open the selected footprint documentation pdf file using the default pdf viewer. |
| | Enable or disable the filtering to limit the list of footprints to the footprint filters of the selected component. |
| | Enable or disable the filtering to limit the list of footprints using the pin count of the selected component. |
| | Enable or disable filtering to limit the list of footprints using the selected library. |

## 4.3  Main Window Keyboard Commands

The following table lists the keyboard commands for the main window:

| | |
|---|---|
| Right Arrow / Tab | Activate the next pane to the right of the currently activated pane. Wrap around to the first pane if the last pane is currently activated. |

| Left Arrow | Activate the next pane to the left of the currently activated pane. Wrap around to the last pane if the first pane is currently activated. |
|---|---|
| Up Arrow | Select the previous item of the currently selected list. |
| Down Arrow | Select the next item of the currently selected list. |
| Page Up | Select the item up one full page of the currently selected list. |
| Page Down | Select the item down one full page of the currently selected list. |
| Home | Select the first item of the currently selected list. |
| End | Select the last item of the currently selected list. |

## 4.4   CvPcb Configuration

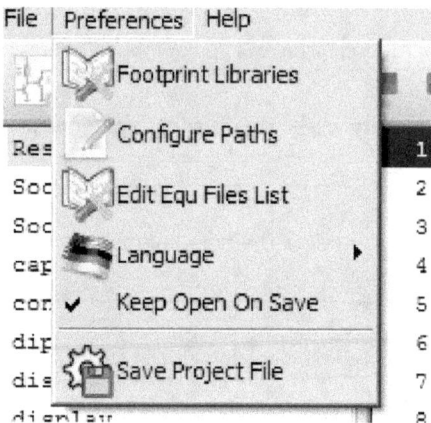

CvPcb can be automatically closed after saving the footprint association file, or not.

Invoking the "Libraries" entry in the "Preferences" menu displays the library configuration dialog.

Depending on the CvPcb version, there are 2 different methods of library management:

- The legacy management, using *.mod files, and a library list of files.

- The new "Pretty" format, using one file by footprint. It uses a folder list. Each folder (*.pretty folder name a library. When using this new method of library management, You can also use native libraries originating fi GEDA/GPCB or even Eagle xml format files.

# 5 Footprint Libraries Management

## 5.1 Important remark:

*This section is relevant only for KiCad versions since December 2013*

## 5.2 Footprint Library tables

Since December 2013, Pcbnew and CvPcb uses a new library management tool based on ***footprint library tal*** which allows **direct use of footprint libraries** from

- KiCad Legacy footprint libraries (.mod files)

- KiCad New *.pretty* footprint libraries (on your local disk) (folders with .pretty extension, containing .kicad_r files)

- KiCad New *.pretty* footprint libraries (on our Github server, or other Github server)

- GEDA libraries (folders containing .fp files)

- Eagle footprint libraries

---

**Note**

- you can write only KiCad *.pretty* footprint library folders on your local disk (and the .kicad_mod files inside th folders).

- All other formats are read only.

---

The image below shows the footprint library table editing dialog which can be opened by invoking the "Footp Libraries" entry from the "Preferences" menu.

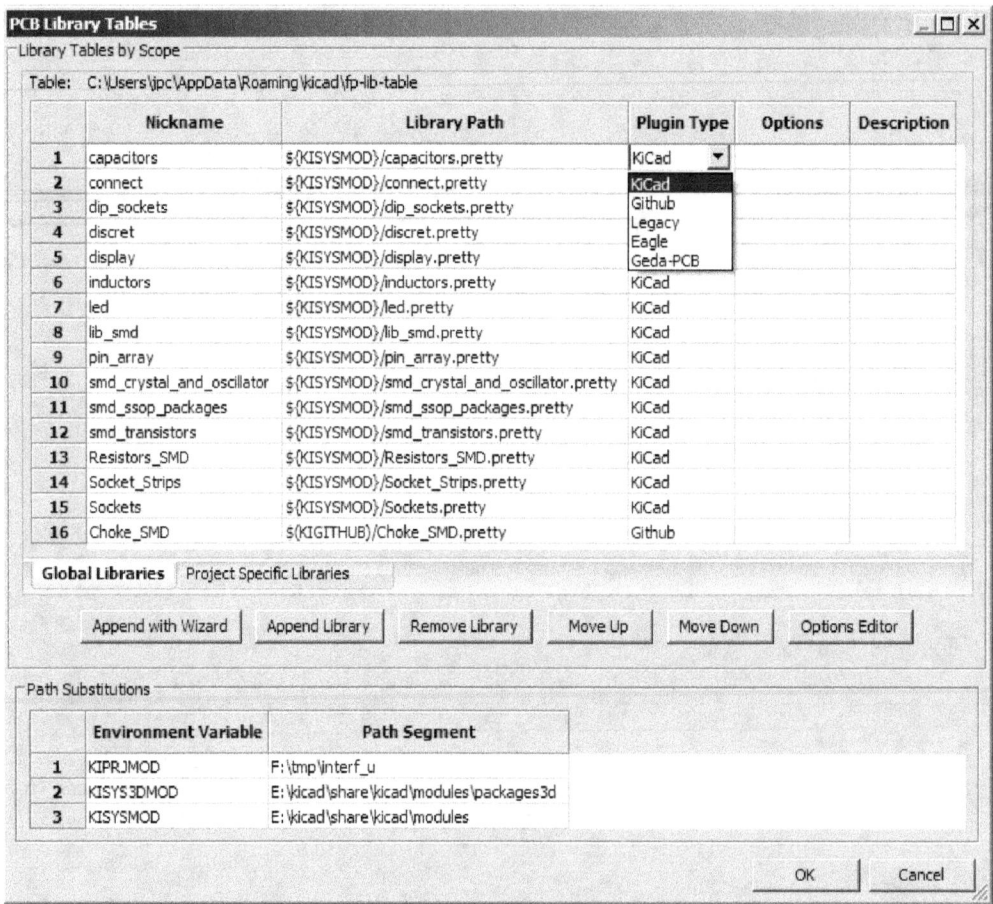

The footprint library table is used to map a footprint library of any supported library type to a library nickna **This nickname is used to look up footprints** instead of the previous method which depended on library sea path ordering.

This allows CvPcb to access footprints with the same name in different libraries by ensuring that the correct footp is loaded from the appropriate library. It also allows CvPcb to support loading libraries from different PCB edi such as Eagle and GEDA.

### 5.2.1  Global Footprint Library Table

The global footprint library table contains the list of libraries that are always available regardless of the currei loaded project file. The table is saved in the file fp-lib-table in the user's home folder. The location of this folde dependent upon the operating system being used.

### 5.2.2  Project Specific Footprint Library Table

The project specific footprint library table contains the list of libraries that are available specifically for the currei loaded project file. The project specific footprint library table can only be edited when it is loaded along with project netlist file. If no project file is loaded or there is no footprint library table file in the project path, an em table is created which can be edited and later saved along with the footprint assignment file.

### 5.2.3 Initial Configuration

The first time Pcbnew or CvPcb is run and the global footprint table file **fp-lib-table** is not found in the user home folder, Pcbnew or CvPcb will attempt to copy the default footprint table file fp-lib-table stored in the syst s KiCad template folder to the file fp-lib-table in the user' s home folder.

If fp-lib-table cannot be found, an empty footprint library table will be created in the user' s home folder. If happens, the user can either copy fp-lib-table manually or configure the table by hand.

The default footprint library table includes many of the standard footprint libraries that are installed as part of KiC

Obviously, the **first thing** to do is to modify this table (add/remove entries) according to your work and the libra you need for all your projects.

(Too many libraries to load is time consuming)

### 5.2.4 Adding Table Entries

In order to use a footprint library, it must first be added to either the global table or the project specific table. ' project specific table is only applicable when you have a net list file open.

**Each library entry must have a unique nickname.**

This does not have to be related in any way to the actual library file name or path. The colon : character cannot used anywhere in the nickname. Each library entry must have a valid path and/or file name depending on the t of library. Paths can be defined as absolute, relative, or by environment variable substitution (see section below).

The appropriate plug in type must be selected in order for the library to be properly read. KiCad currently supp reading KiCad legacy, KiCad Pretty, Eagle, and GEDA footprint libraries.

There is also a description field to add a description of the library entry. The option field is not used at this tim adding options will have no effect when loading libraries.

- Please note that you cannot have duplicate library nicknames in the same table. However, you can have duplic library nicknames in both the global and project specific footprint library table.

- The project specific table entry will take precedence over the global table entry when duplicated names oc When entries are defined in the project specific table, an fp-lib-table file containing the entries will be written i the folder of the currently open net list.

### 5.2.5 Environment Variable Substitution

One of the most powerful features of the footprint library table is environment variable substitution. This allows to define custom paths to where your libraries are stored in environment variables. Environment variable substitut is supported by using the syntax `${ENV_VAR_NAME}` in the footprint library path.

By default, at run time KiCad defines **two environment variables**:

- the KIPRJMOD environment variable. This points always the current project directory and cannot be modified.

- the KISYSMOD environment variable. This points to where the default footprint libraries that were installed v KiCad are located.

You can override KISYSMOD by defining it yourself in preferences/Configure Path which allows you to substitute y own libraries in place of the default KiCad footprint libraries.

When a project netlist file is loaded, CvPcb defines the KIPRJMOD using the file path (the project path).

Pcbnew also defines this environment variable when loading a board file.

This allows you to store libraries in the project path without having to define the absolute path (which is not alw known) to the library in the project specific footprint library table.

### 5.2.6  Using the GitHub Plugin

The GitHub is a special plugin that provides an interface for read only access to a remote Git Hub repository consist of pretty (Pretty is name of the KiCad footprint file format) footprints and optionally provides "Copy On Wr (COW) support for editing footprints read from the GitHub repo and saving them locally. Therefore the "Git H plugin is for **read only accessing remote pretty footprint libraries at** https://github.com. To add a Gitł entry to the footprint library table the "Library Path" in the footprint library table row a must be set to a v GitHub URL.

For example:

https://github.com/liftoff-sr/pretty_footprints

or

https://github.com/KiCad

Typicality GitHub URLs take the form:

https://github.com/user_name/repo_name

The "Plugin Type" must be set to "Github". To enable the "Copy On Write" feature the option **allow_pretty_wri** must be added to the "Options" setting of the footprint library table entry. This option is the "Library Path" for l storage of modified copies of footprints read from the GitHub repo. The footprints saved to this path are combi with the read only part of the Git Hub repository to create the footprint library. If this option is missing, then Git Hub library is read only. If the option is present for a Git Hub library, then any writes to this hybrid library go to the local *.pretty directory. Note that the github.com resident portion of this hybrid COW library is alw read only, meaning you cannot delete anything or modify any footprint in the specified Git Hub repository direc The aggregate library type remains "Github" in all further discussions, but it consists of both the local read/w portion and the remote read only portion.

The table below shows a footprint library table entry without the option **allow_pretty_writing_to_this_di**

| Nickname | Library Path | Plugin Type | Options | Descripti |
|----------|--------------|-------------|---------|-----------|
| github | https://<br>github.com/liftoff-sr/pretty_footprints | Github | | Liftoff's<br>GH<br>footprints |

The table below shows a footprint library table entry with the COW option given. Note the use of the environm variable ${HOME} as an example only. The github.pretty directory is located in ${HOME}/pretty/ path. Anyt you use the option **allow_pretty_writing_to_this_dir**, you will need to create that directory manually in adva and it must end with the extension **.pretty**.

| Nickname | Library Path | Plugin Type | Options | Descripti |
|----------|--------------|-------------|---------|-----------|
| github | https://<br>github.com/liftoff-sr/pretty_footprints | Github | allow_pretty_writing_to_this_d<br>${HOME}/pretty/<br>github.pretty | Liftoff's<br>GH<br>footprints |

Footprint loads will always give precedence to the local footprints found in the path given by the option **allow_pret**
Once you have saved a footprint to the COW library's local directory by doing a footprint save in the footprint edi
no Git Hub updates will be seen when loading a footprint with the same name as one for which you've saved loca

Always keep a separate local *.pretty directory for each Git Hub library, never combine them by referring to the sa
directory more than once.

Also, do not use the same COW (*.pretty) directory in a footprint library table entry. This would likely create a m

The value of the option **allow_pretty_writing_to_this_dir** will expand any environment variable using the
notation to create the path in the same way as the "Library Path" setting.

What is the point of COW? It is to turbo-charge the sharing of footprints.

If you periodically email your COW pretty footprint modifications to the GitHub repository maintainer, you can l
update the Git Hub copy. Simply email the individual *.kicad_mod files you find in your COW directories to
maintainer of the GitHub repository. After you have received confirmation that your changes have been commit
you can safely delete your COW file(s) and the updated footprint from the read only part of Git Hub library will l
down. Your goal should be to keep the COW file set as small as possible by contributing frequently to the sha
master copies at https://github.com.

### 5.2.7  Usage Patterns

Footprint libraries can be defined either globally or specifically to the currently loaded project. Footprint libra
defined in the user's global table are always available and are stored in the fp-lib-table file in the user's home fol

Global footprint libraries can always be accessed even when there is no project net list file opened.

The project specific footprint table is active only for the currently open net list file.

The project specific footprint library table is saved in the file fp-lib-table in the path of the currently open net li
You are free to define libraries in either table.

There are advantages and disadvantages to each method. You can define all of your libraries in the global table wl means they will always be available when you need them. The disadvantage of this is that you may have to sea through a lot of libraries to find the footprint you are looking for. You can define all your libraries on a project spec basis.

The advantage of this is that you only need to define the libraries you actually need for the project which cuts d on searching.

The disadvantage is that you always have to remember to add each footprint library that you need for every proj You can also define footprint libraries both globally and project specifically.

One usage pattern would be to define your most commonly used libraries globally and the library only require for project in the project specific library table. There is no restriction on how you define your libraries.

## 5.3  Using the Footprint Library Table Wizard

A wizard to add footprint libraries to the footprint library tables is available from the *footprint library table edi dialog*.

Note also libraries can be any type of footprint library supported by KiCad.

It can add "local" libraries or libraries from a Github repository.

When libraries are on a Github repository, they can be added as remote libraries, or **downloaded and added** *local libraries*.

Here, the local libraries option is selected.

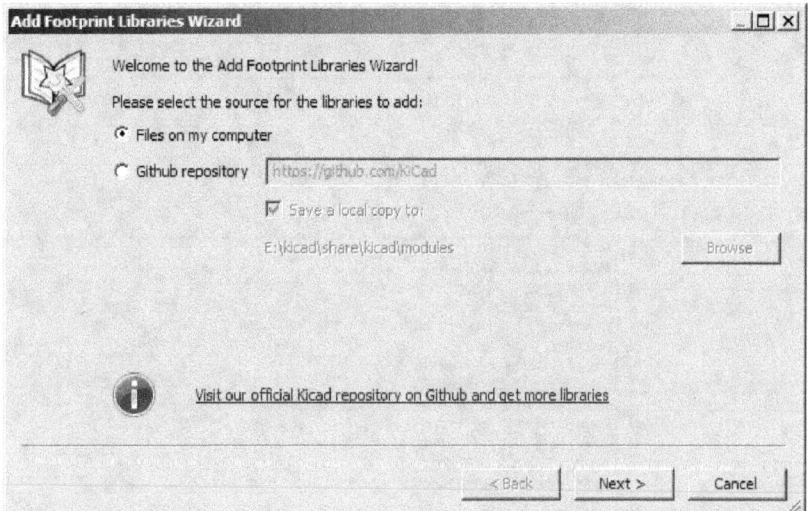

Here, the remote libraries option is selected.

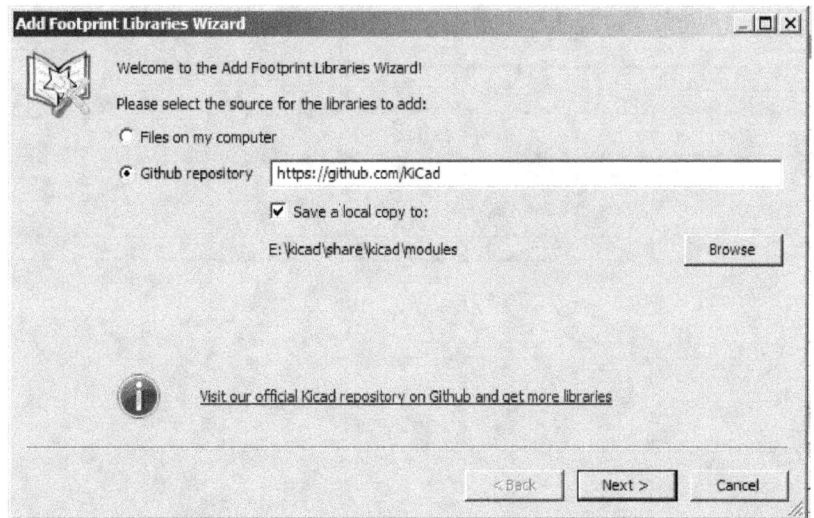

Depending on the selected option, one of these pages will be displayed, to select a list of libraries:

Here, the local libraries option was selected.

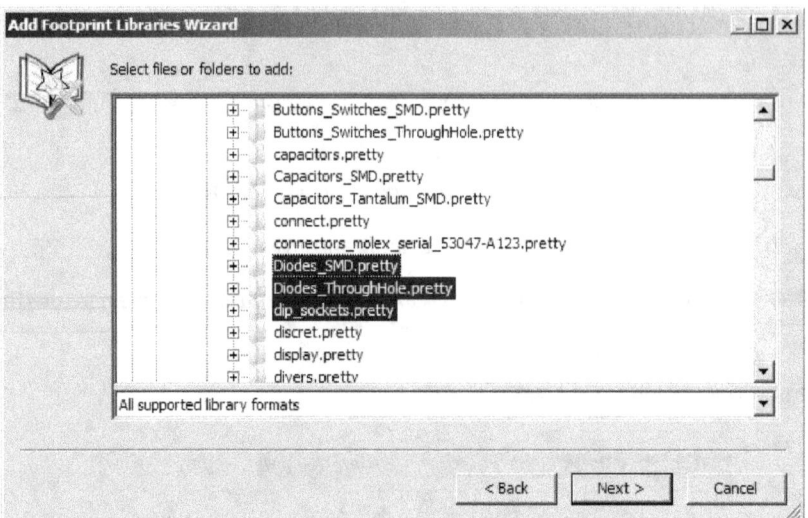

Here, the remote libraries option was selected.

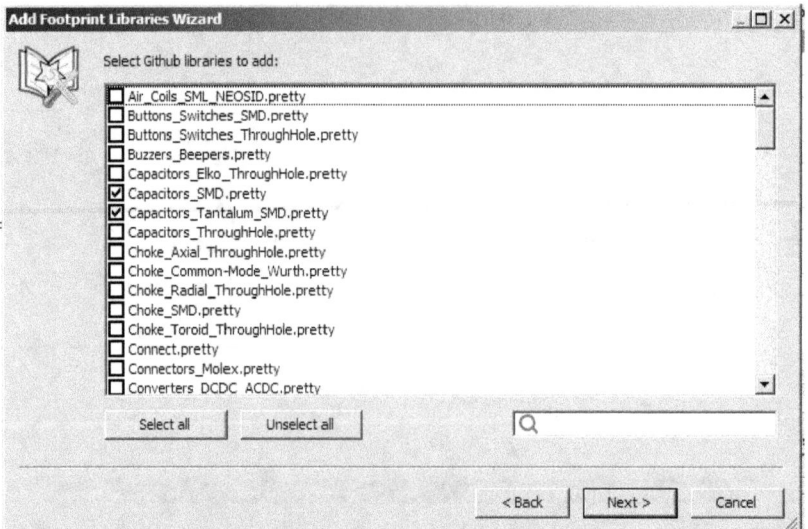

After a set of libraries is selected, the next page validates the choice:

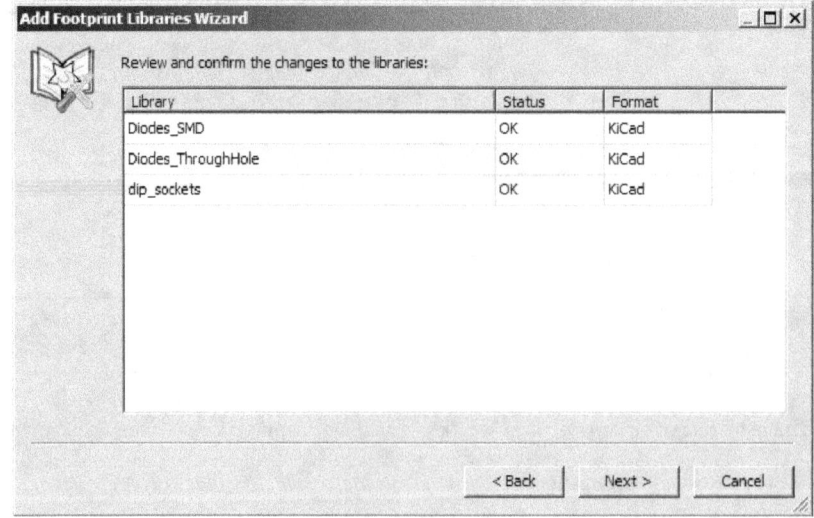

If some selected libraries are incorrect (not supported, not a footprint library ⋯) they will be flagged as "INVAL

.

The last choice is the footprint library table to populate:

- the global table

- the local table (the project specific table)

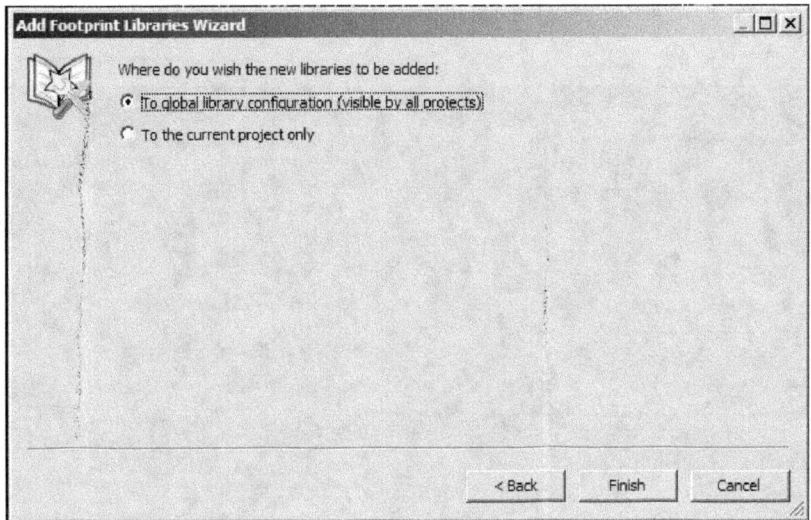

# 6  Viewing the Current Footprint

## 6.1  The view footprint command

The view footprint command displays the footprint currently selected in the *footprint* window. A 3D model of component can be shown if it has been created and assigned to the footprint. Below is the footprint viewer windo

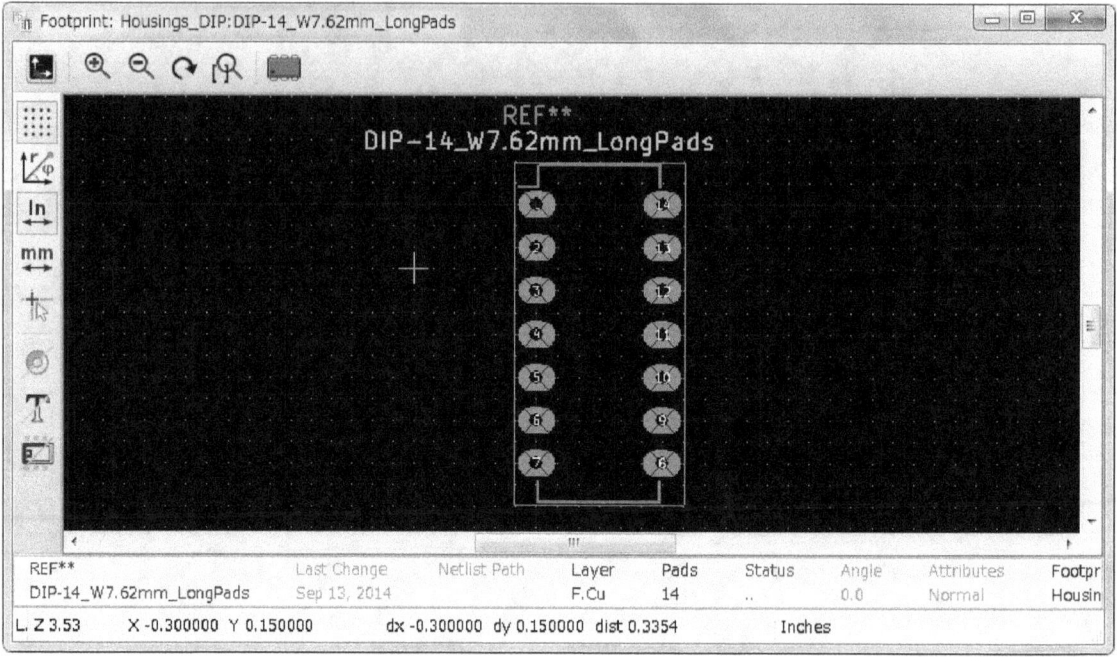

### 6.1.1  Status Bar Information

The status bar is located at the bottom of the CvPcb new main window and provides useful information to the u The following table defines the contents of each pane in the status bar.

### 6.1.2  Keyboard Commands

| | |
|---|---|
| F1 | Zoom In |
| F2 | Zoom Out |
| F3 | Refresh Display |
| F4 | Move cursor to center of display window |
| Home | Fit footprint into display window |
| Space Bar | Set relative coordinates to the current cursor position |
| Right Arrow | Move cursor right one grid position |
| Left Arrow | Move cursor left one grid position |
| Up Arrow | Move cursor up one grid position |
| Down Arrow | Move cursor down one grid position |

### 6.1.3  Mouse Commands

| Scroll Wheel | Zoom in and out at the current cursor position |
|---|---|
| Ctrl + Scroll Wheel | Pan right and left |
| Shift + Scroll Wheel | Pan up and down |
| Right Button Click | Open context menu |

### 6.1.4  Context Menu

Displayed by right-clicking the mouse:

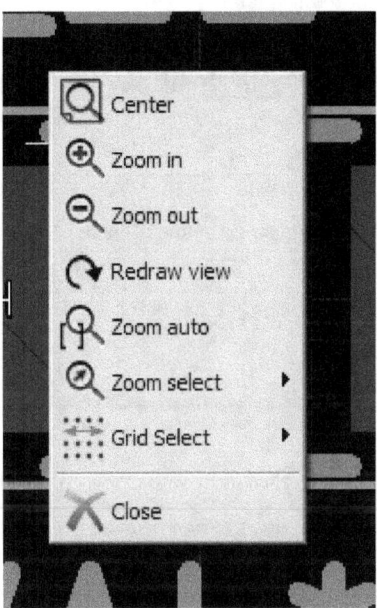

| Zoom Selection (Select Zoom) | Direct selection of the display zoom. |
|---|---|
| Grid Selection (Grid Select) | Direct selection of the grid. |

### 6.1.5  Horizontal Toolbar

| | |
|---|---|
| | Show display options dialog |
| | Zoom in |
| | Zoom out |
| | Redraw |
| | Fit drawing in display area |

Open 3D model viewer

### 6.1.6 Vertical Toolbar

| | |
|---|---|
| ⠿ | Show or hide the grid |
| r/φ | Show coordinates in polar or rectangular notation |
| In | Display coordinates in inches |
| mm | Display coordinates in millimeters |
| ✛ | Toggle pointer style |
| ⊘ | Toggle between drawing pads in sketch or normal mode |
| T | Toggle between drawing text in sketch or normal mode |
| ▣ | Toggle between drawing edges in sketch or normal mode |

## 6.2 Viewing the Current 3D Model

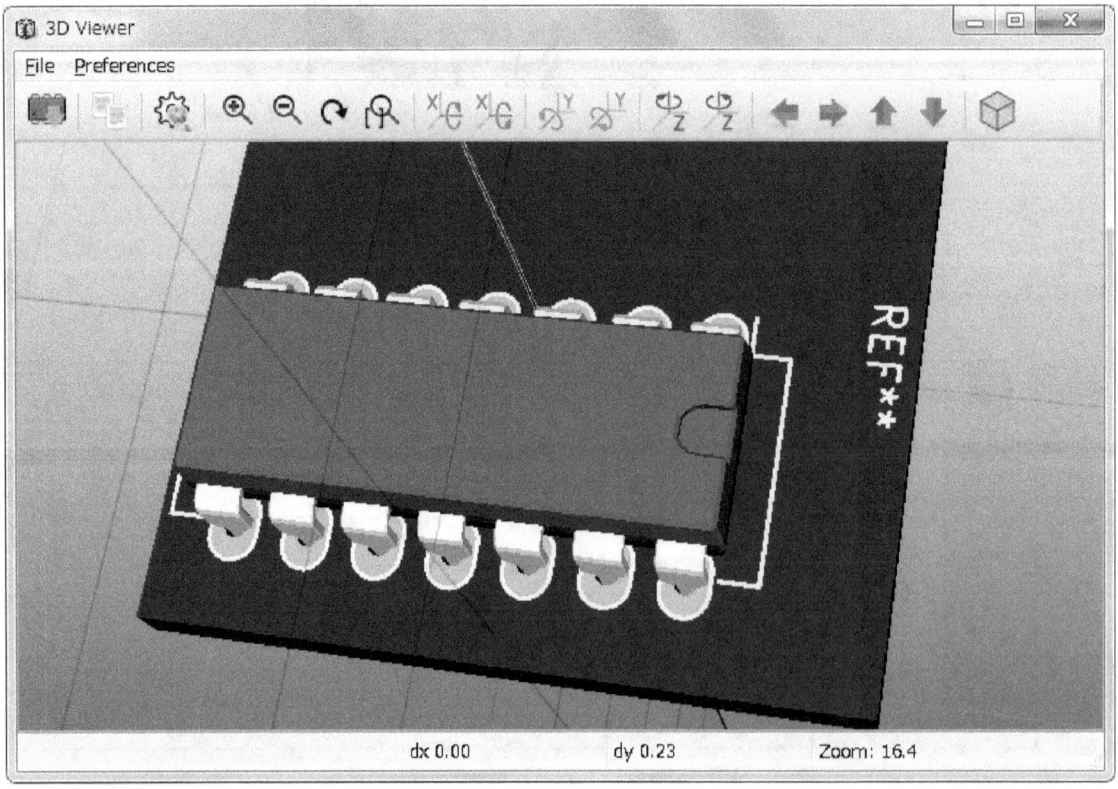

### 6.2.1  Mouse Commands

| Scroll Wheel | Zoom in and out at the current cursor position |
|---|---|
| Ctrl + Scroll Wheel | Pan right and left |
| Shift + Scroll Wheel | Pan up and down |

### 6.2.2  Horizontal Toolbar

| | |
|---|---|
| | Reload the 3D model |
| | Copy 3D image to clipboard |
| | Set 3D viewer options |
| | Zoom in |
| | Zoom out |
| | Redraw |
| | Fit drawing in display area |
| | Rotate backward along the X axis |
| | Rotate forward along the X axis |
| | Rotate backward along the Y axis |
| | Rotate forward along the Y axis |
| | Rotate backward along the Z axis |
| | Rotate forward along the Z axis |
| | Pan left |
| | Pan right |
| | Pan up |
| | Pan down |
| | Toggle orthographic projection mode on and off |

## 7    Using CvPcb to Associate Components with Footprints

### 7.1    Manually Associating Footprints with Components

To manually associate a footprint with a component first select a component in the component pane. Then sele
footprint in the footprint pane by double-clicking the left mouse button on the name of the desired footprint. '
unassigned next component in the list is automatically selected. Changing the component footprint is performe
the same manner.

### 7.2    Filtering the Footprint List

If the selected component and/or library is highlighted when the one or more of the filtering option is enabled,
displayed footprint list in CvPcb is filtered accordingly.

The icons enable and disable the filtering feature. When the filtering is not enabled, the
footprint list is shown.

Without filtering:

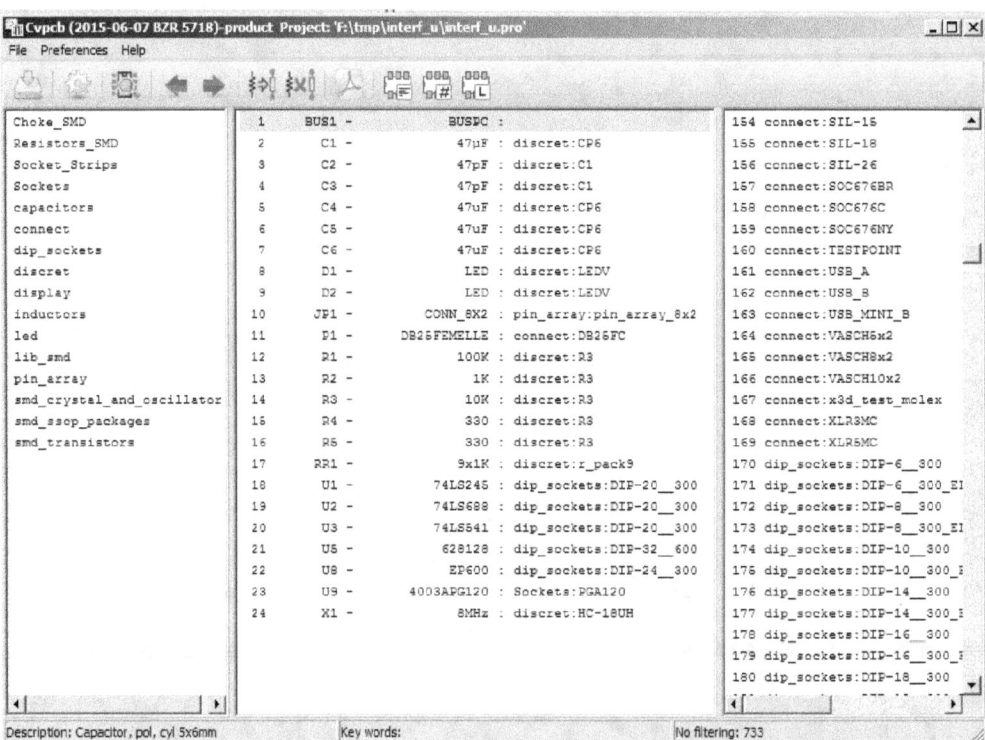

Filtered by list of footprint filters assigned to the selected component. The component filters are listed on the ce
pane of the status bar at the bottom of the main window.

Filtered by the footprint filter of the selected component:

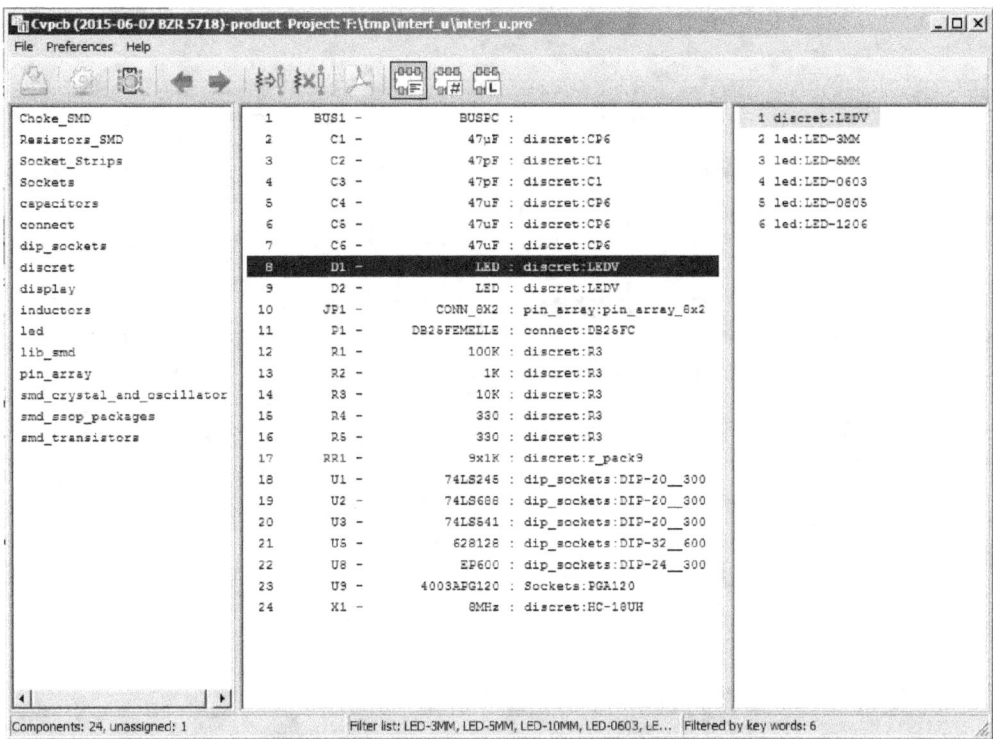

In the component library editor in Eeschema, the footprint list was set using the entries in the footprint filter tal
the component properties dialog as shown below.

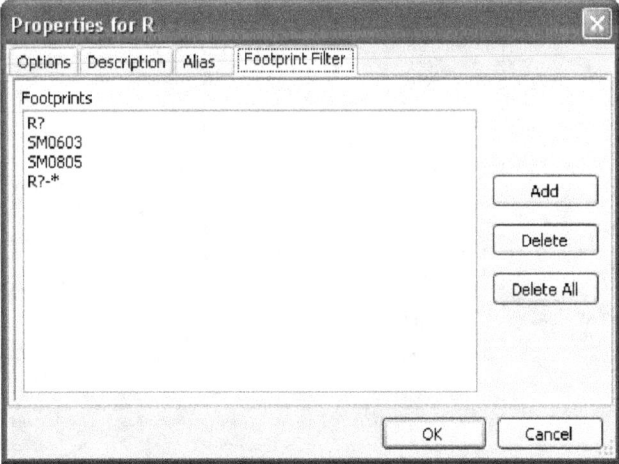

Filtered by the pin count of the selected component:

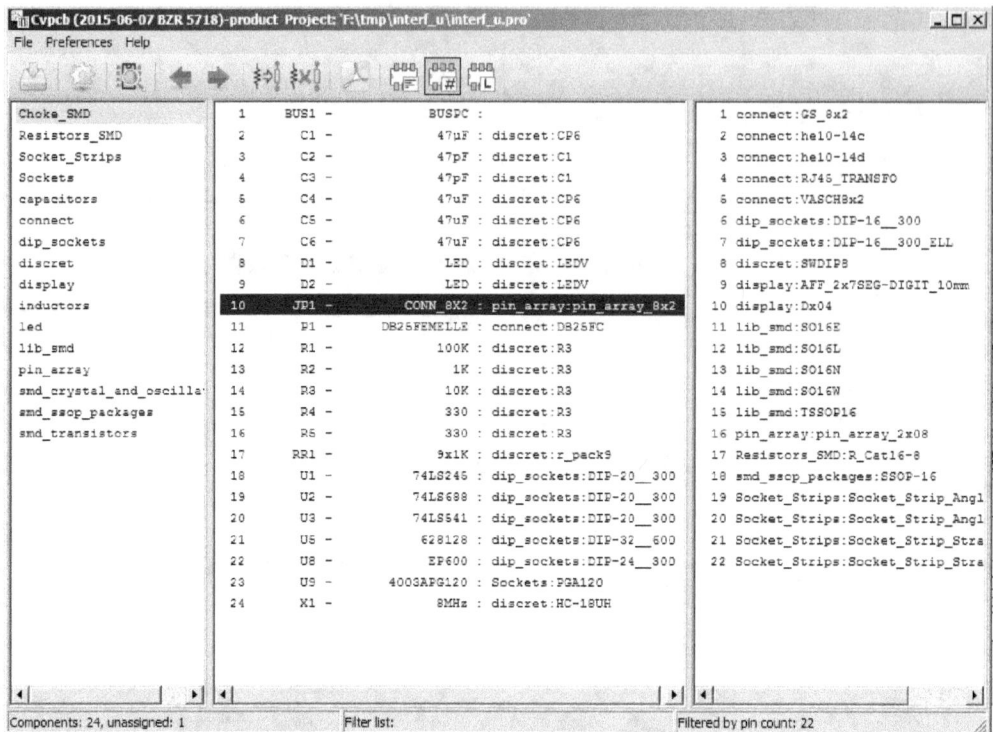

Filtered by the selected library.

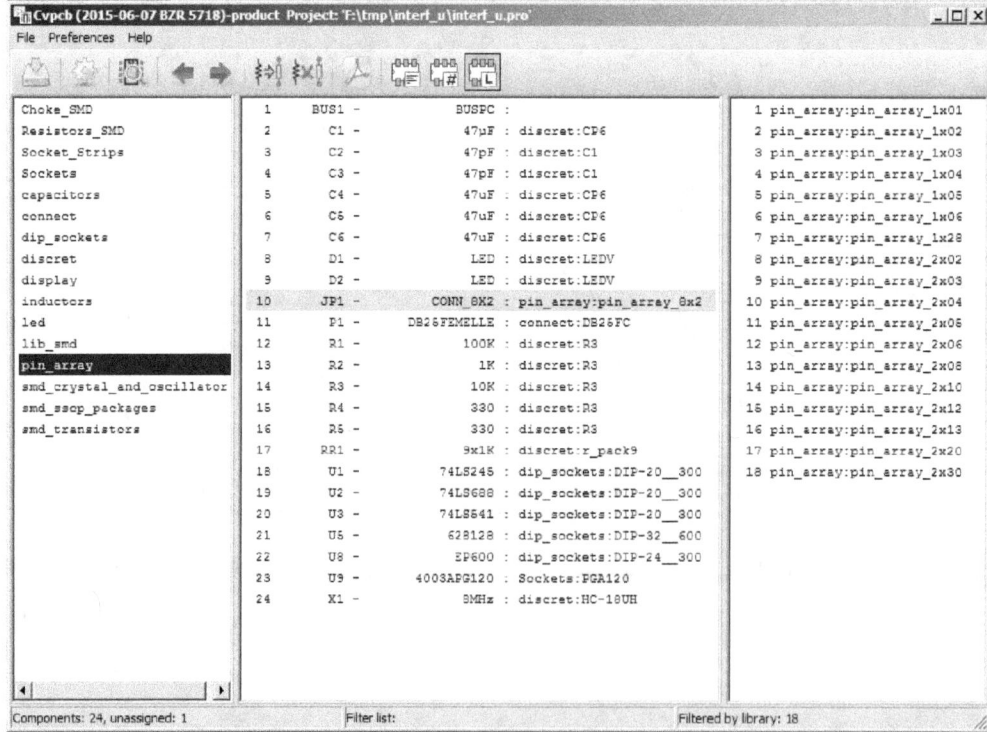

The filtering can be combined to form more complex filtering to help reduce the number of footprints in the footp

pane.

Filtered by the selected component pin count and the component filter:

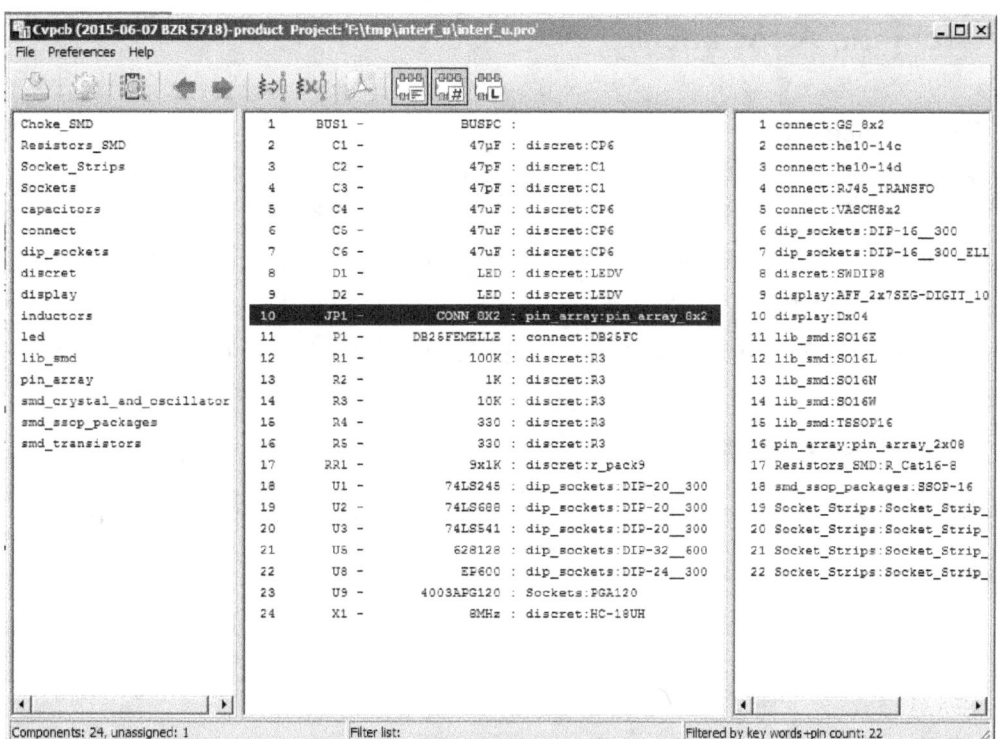

# 8  Automatic Associations

## 8.1  Equivalence files

Equivalence files allow for automatic assignment of footprints to components.

They list the name of the corresponding footprint according to the name (*value field*) of the component. These :
typically have the **.equ** file extension.

They are plain text files and may be edited by using any plain text editor, and must be created by the user.

## 8.2  Equivalence File Format

Equivalence files consist of one line for each component. Each line has the following structure:

**'component value'    'footprint name'**

Each name must be single quoted by the ' character and the component and footprint names must be separated
one or more spaces.

*Example:*

If the U3 component is circuit 14011 and its footprint is 14DIP300, the line is:

**'14011'    '14DIP300'**

Any line starting with # is a comment.

Here is an example equivalence file:

```
#integrated circuits (smd):
'74LV14' 'SO14E'
'74HCT541M' 'SO20L'
'EL7242C' 'SO8E'
'DS1302N' 'SO8E'
'XRC3064' 'VQFP44'
'LM324N' 'SO14E'
'LT3430' 'SSOP17'
'LM358' 'SO8E'
'LTC1878' 'MSOP8'
'24LC512I/SM' 'SO8E'
'LM2903M' 'SO8E'
'LT1129_SO8' 'SO8E'
'LT1129CS8-3.3' 'SO8E'
'LT1129CS8' 'SO8E'
'LM358M' 'SO8E'
'TL7702BID' 'SO8E'
'TL7702BCD' 'SO8E'
'U2270B' 'SO16E'
#Xilinx
'XC3S400PQ208' 'PQFP208'
```

```
'XCR3128-VQ100' 'VQFP100'
'XCF08P' 'BGA48'

#upro
'MCF5213-LQFP100' 'VQFP100'

#regulators
'LP2985LV' 'SOT23-5'
```

## 8.3  Automatically Associating Footprints to Components

Click on the automatic footprint association button on the top toolbar to process an equivalence file.

*All components found by their value in the selected equivalence (*.equ) file will have their footprint automatic assigned.*